EARTH'S ALIEN SYLLABUS

Martin Thomas

Earth's Alien Syllabus

The moral rights of the author have been asserted by them.

© Martin Thomas 2025

10 9 8 7 6 5 4 3 2 1

All rights reserved

No part of this publication may be reproduced, stored in a retrieval system, or transmitted in any form or by any other means (electronic, mechanical, photocopying, recording or otherwise) or used to train any artificial intelligence technologies without the prior written permission of the author. Subject to EU law, the author expressly reserves this work from the text and data mining exemption.

Previous publication by the author:

UFO – Friend or Foe?   August 2025   978-1-918045-21-5

We Own 29% - ET Has the Rest   September 2025   978-1-83709-256-7

Web Address: MartinThomasAuthor.com

Earth's Alien Syllabus

Dedication

This work is dedicated to the BLAZE Television Channel for making me think, to Google for helping me find information and to Wikipedia for denying everything.

The Scientific Method

Develop an hypothesis which can be used to make testable predictions. Test these predictions and if they work, test again for repeatability.

The Engineering Method

If it walks like a duck, and quacks like a duck, it probably is a duck.

Earth's Alien Syllabus

Earth's Alien Syllabus

Table of Contents

		Preface	vii
1		**THE 50,000 YEAR PLAN**	1
	a)	I Get Convinced UAPs Are Real	1
	b)	The Anomaly	2
2		**40,000 YEARS LATER**	4
	a)	Agriculture	4
	b)	Building Stone Structures	5
	c)	Metallurgy, Mining & Blacksmithing	7
	d)	The Greek Classical Period	9
	e)	China & Central America	10
3		**THE LAST THOUSAND YEARS**	13
	a)	The Renaissance	13
	b)	The Railways	15
	c)	Chemistry	16
	d)	Electrical Power Generation & Transmission	18
	e)	Telecommunications	20
	f)	Chinese Development	21
4		**THE LAST CENTURY**	23
	a)	The Development of US Society	23
	b)	Power Transmission	24
	c)	Extra-Terrestrial Technology	25
	d)	The "Other" Team	27
	e)	An Alternative Syllabus	28
	f)	The Majestic Twelve	29
	g)	Current United States' Activities	31
5		**THE WAY FORWARD**	35
	a)	Spaceflight	35
	b)	Trading with "Other" Colonies	36
		Appendices	38
		Index	67
		References	71

Earth's Alien Syllabus

PREFACE

In my first two books "UFO – Friend or Foe?[1]" and "We Own 29% - ET Has the Rest" [2] I have made a broad-ranging sweep through the history of UAP and "Other" sightings across as much of our planet as I could.

The terms Unidentified Flying Object (UFO) and Unidentified Submersible Object (USO) have been merged as Unidentified Anomalous Phenomena (UAPs). I will be using the term UAP throughout this book.

The term "Alien" has been used to describe any occupants or manufacturers of UAPs. However, this word has many pejorative associations – it means basically "not one of us" and is frequently used to describe immigrants, whose presence may be illegal, or who are disliked because of their race, colour or creed. I intend to use the word "Others" throughout this book - perhaps not an ideal term, but at least it avoids many of the pre-assumptions inherent in traditional terms.

In reviewing evidence of all the sightings of UAPs, I was easily able to convince myself, and hopefully my readers, that they really existed, and that they had some bases underground, and underwater.

President Dwight Eisenhower made a grave mistake when he signed a deal with the Small Grays where, in return for information about "Other" technology, he allowed them a base in the US and granted them permission to take cattle and humans for experimental purposes, provided that the humans were returned unharmed, and they limited the numbers that they took. The Small Grays, once they were comfortably installed, have broken every part of the agreement, and the US can do nothing about it. In addition they have stopped the US (and Possibly Russia) from keeping their nuclear weapons at readiness, and have insisted on their presence being kept secret until they authorise disclosure.

There is evidence that a group of "Others" have been nurturing the human race for tens of thousands of years, and they did warn the US against signing a deal with the Small Grays.

In my last book[3], I showed that there are a number of "Other" races who have established small colonies on Earth, together with many bases. Some of these may be the same "Others" who have been acting to encourage the development of the Human Race. It will be interesting to see whether they come into conflict with the US Military-Industrial entity known as Majestic-12.

I showed that Majestic-12 was developing spacecraft, but it is difficult to determine how advanced these are, or the uses to which they have already been put.

Chapter 1

THE 50,000 YEAR PLAN

(You Can Forget Those 5-year Plans)

a) <u>I GET CONVINCED UAPs ARE REAL</u>

Ancient clay tablets[4] found in Sumeria dating back thousands of years suggest that Humans were visited by strangers called the Anunnaki, a race of giants that they worshipped as Gods. When viewed with a modern eye, these would probably be considered as some form of Extra-Terrestrial.

In my first book, I reviewed a wide range of UAP sightings to try and get a feel for their likely veracity. I soon excluded any sightings where there was only one witness on the basis that I was not able to go into their possible motivations, whether honest, dishonest or delusional. Then, I eliminated mysterious cloud formations and single unmoving points of light, on the basis that they could so easily be miss-identifications.

What remained was a group of sightings which could not be explained easily. If it wasn't so worrying, it would be comical: there were desperate attempts by officialdom to account for these, including such fantasies as marsh gas burning[5] high in the air rather than getting diluted as it rose, or the highly improbable chance coincidence[6] of a meteor and an earthquake in a non-earthquake zone.

Then we have threats on witnesses and their families, either of violence or of destruction of their careers.

Even if the evidence of these remaining sightings were not so convincing, the desperate attempts at denial would probably have convinced me of the reality of UAPs and "Others".

I remain certain that UAPs and "Others" exist, even if they have been in our folklore as far back as human memory can stretch, and I hope that I managed to convince some of my readers in my first book. I accept that there are sincere skeptics who cannot believe, but I invite them to read further to see whether some strange events are better explained in my interpretation, than in the "Official" version. Otherwise, just read and enjoy it as a fantasy.

b) <u>THE ANOMALY</u>

Some Geneticists say that the human race has developed faster than other animals in the period since the extinction of the dinosaurs. For a long time, they all developed at the same rate then, suddenly, about 50,000 years ago, there was a sudden sprint in human brain development, which may have, in turn, triggered an explosion of creativity and invention.

In 2006, geneticist Dr Coleen Clements from Rochester University, published a book[7] showing that, at some time in history between 60,000 and 14,000 years ago, something called the "D allele" entered the human microcephalin gene of our DNA[8]. This had the effect of encouraging the growth and development of the human brain[9].

The usual way in which this might happen is by interbreeding, but tests[10] have shown that neither of the possible candidate hominids, Neanderthals or Denisovians, had this either, so it must have come from somewhere else. She proposed that it was introduced by "Others". This was controversial, but has not been disproved.

If any of our friendly skeptics are still with me, the challenge is for you to think up an alternative.

It is suggested that a race called the Anunnaki[11] were responsible for this modification to our genes. They had discovered an inquisitive ape on this planet, ranging across the whole world which, unlike Neanderthal, Flores, Denisovian, or other strains of hominid around at the time, they considered could be developed into a contributing member of their confederation. If true, this plan reshapes our whole understanding of humanity's purpose.

Chapter 2

40,000 YEARS LATER

a) <u>THE GRAND PLAN</u>

It seems amazing to us that the Anunnaki thought so long-term. About 10.000 years ago, the Human Race had developed to the level where they could commence the next stage of their work. Many tribes around the world have folklore describing how "Gods" came to teach them. This happened in North, Central and South America, Europe, India, Australia and New Zealand. The descriptions of the "Gods" vary, but they were all doing the same thing. Seen with modern eyes, these "Gods" were "Others".

b) <u>AGRICULTURE</u>

About 12,000 years ago, in Mesopotamia for example, it all started with a human tribe of hunter-gatherers which was eking out a living on the banks of the Euphrates. Clay tablets[12], found by modern archaeologists, describe how a species called the Anunnaki, giant humanoids, possibly assisted by Avians[13], a bird-like species, arrived there and started introducing them to ways of making their lives easier. The humans learnt agriculture, mathematics, astronomy, writing, schooling and the calendar. They were the first to introduce the 24 hour day and the 60 minute hour. It was also during this time that the wheel was invented.

All these innovations are important, but that of agriculture was crucial. By learning about irrigation, cattle husbandry, the gathering and sowing of seeds and rotation of crops, they could produce more food for the same effort. Alternatively, it provided spare time for other activities.

Pictures of Anunnaki and Avians appear in Egyptian paintings and carvings, and this suggests that they were spreading their tuition widely. They would probably have had dealings with the early Persians, Turks and Greeks as well.

It is recorded that the Anunnaki left Earth about 5000 years ago, by which time all these innovations were spreading around the world.

c) BUILDING STONE STRUCTURES

One of the more contentious points in archaeology is whether early civilizations such as the Mayans or Egyptians had the capability to cut and move these massive stones, and erect these colossal edifices. It would certainly have been easier for them if they had received the advice or assistance of "Others".

Tales of the construction of Stonehenge describe the stones being moved by magic (or music?). The islanders of Pohnpei in Micronesia say the same about Nan Madol. In South America, it has been said that the ancients had a way of softening rock to shape it.

We'll never know, but this could have been the first step in creating our civilization. Certainly someone like the "Other" race known as Dwarfs, small people famed for their stone and metal working skills, (See Chapter3 section d) would have had the capability to work with edifices like these, and it is possible that they did exactly that.

The question is twofold –

- Who built these monolithic structures?
- Why?

Pyramids have turned up all over the world, but we don't know whether they were the result of parallel development, or simply copies, assuming humans from Central America were capable of visiting Egypt at that time. Examples are:

- Egypt – obviously   2650 BCE
- Baghdad[14]   1500 BCE
- China[15]   2300 BCE
- Central America[16]   900 BCE

It would strange if the Star Trek's "Hodgkin's Law of Parallel Planetary Development" was actually applicable!

I am left wondering whether the idea, and possibly the design, for these comes from "Others" rather than humans, even if the execution was purely human.

It is the spread of pyramid sites around the world which is my challenge to the skeptics here.

d) METALLURGY, MINING & BLACKSMITHING

Once humans had spare time from scraping together enough food to survive, ideas could be introduced to encourage their curiosity and creative urges.

Copper had been worked for a long time during the Stone Age but, about 3,000 BCE (5,000 years ago by coincidence), tin was first discovered in Anatolia in Turkey[17]. Very soon after that, tin was added to copper to make bronze, and the Bronze Age began[18], again in Anatolia. Who thought to do that? It certainly advanced civilization significantly.

It was not long before the demand for tin outstripped the supply around the Mediterranean[19], and skilled miners, most probably Dwarfs, started looking for new sites, heading westward to Spain and Portugal, northwards to Britain and Germany, and eastwards to India, passing on their knowledge and craft to the locals and thereby earning their reputation which lasts to this day. Appendix 4 contains information on Dwarfs. The Bronze Age started in Britain in about 2,000 BCE.

However, the Anatolians were not resting on their laurels. In 1,300 BCE, they were the first to produce iron[20], from ore as opposed to using meteoritic iron, initiating the Iron Age, and gradually bringing the Bronze Age to an end. This happened in Britain in about 800 BCE.

The Anatolians were clearly feeling unappreciated so they became the first to produce high carbon steel[21]. Even into the Common Era, the manufacture of Damascus steel required skilled forging technique to produce very high quality swords. This was invented in India, perhaps by Dwarfs working there.

Toledo sword manufacturing in Spain, during the same period, used the same forging principles as for Damascus steel, but the changes in weaponry proved the death knell for many of the artisan manufacturers[22]. However Spain still has a flourishing steel industry with a reputation for fine knives. There were indications of a Dwarf base in the north[23], near Barcelona or the Basque country, where the first Spanish iron and steel were produced, but they appear to have left in the 1960s.

In Pittsburg in the United States, there were a number of multiple sightings of Dwarfs between 1905 and 1956[24]. Since the 1960s, there have been many claims of Small Gray abductions, and many sightings of UAPs and of Bigfoot, but only one of a Dwarf in 1965. The city is close enough to Lake Erie for there to have been an "Other" base there, or they could have set up a now-abandoned base in a nearby lake.

In Northumberland in England, lived the Simonside Dwarfs[25] also known as Brownmen, Bogles, or Duergar. They appear to have moved away in the 1970s, judging from the sighting reports[26]. One of their parting shots in guiding the human race in metal-working may have been in Sheffield in Britain in 1856, in inspiring Sir Henry Bessimer's[27] nighttime dreams to come up with his process of blowing air through molten iron to produce steel.

What has happened is that a small group of miners, metallurgists and blacksmiths has taken civilization forward in leaps and bounds.

It is evident, though; that the Dwarfs consider their job mainly completed, and have withdrawn from many of their old bases. Still, it is interesting to see that the Dwarf colonies are near the centre of Italian steel-making, and near Sweden's furnaces. Only in Brazil and Argentina do they seem still to be active.

e) THE GREEK CLASSICAL PERIOD

Between the dates 624 BCE to 270 BCE, the Greeks produced some of the greatest minds of the period:

- Hales (c. 624–546 BCE): Often considered the first philosopher, known for his astronomical observations and ideas about water as the fundamental substance.
- Anaximander (c. 610–546 BCE): A Milesian philosopher who proposed the apeiron (the indefinite or boundless) as the origin of all things.
- Pythagoras (c. 570–495 BCE): Known for his contributions to mathematics and his mystical philosophy centered on the transmigration of souls.
- Socrates (c. 470–399 BCE): Known for emphasizing virtue and the use of dialectic to pursue truth.
- Plato (c. 427–347 BCE): Socrates' student, famous for his theory of Forms and advocating for a polity governed by philosophers.
- Aristotle (384–322 BCE): Plato's most famous student, a polymath and founder of the Peripatetic school, whose writings span diverse subjects.
- Diogenes (c. 404–323 BCE): A leading figure in the development of Cynicism.

- Zeno (c. 334–262 BCE): Founder of Stoicism, a philosophy focused on achieving tranquility by accepting what one cannot control
- Pyrrho (c. 360–270 BCE): Credited as the founder of Skepticism and Pyrrhonism.

In part, this was due to Greece being a slave society, with Athens having about one third of its population as slaves. This gave the Citizens the time to indulge themselves.

It is not clear how much influence "Others" had in bringing this about and, if so, how did they do it?

Obviously this is a case where the skeptic is just as likely to be right as anyone else.

f) CHINA & CENTRAL AMERICA

So far, I have been dealing with what is termed Western Civilization, but that is not the only one. Many human races' folklores tell of Gods visiting from the sky and teaching them the fundamentals of agriculture. Few of these progressed further.

The Mayans advanced into constructing massive stone buildings, but remained in the Stone Age.

The Chinese Bronze Age started in approximately 2000 BCE and, as the Chinese starting using bronze at roughly the same time as Britain, they could have been the recipients of knowledge from the same Anatolian source. They entered the Iron Age at roughly 500 BCE, again approximately matching the British timescale.

Chinese philosophers[28] emerged during 770–220 BCE, a period of intellectual flourishing known as the Hundred Schools of Thought, when they developed foundational concepts in Confucianism, Daoism, Mohism, and Legalism.

- Confucius (551–479 BCE),
- Laozi (c. 500 BCE),
- Mozi (c. 470–390 BCE),
- Mencius (372–289 BCE),
- Zhuangzi (c. 4th century BCE),
- Han Feizi (died 233 BCE).

These dates tie in with those of the Greek philosophers, so perhaps things were starting to develop in parallel.

This coincided with the rise of the first emperor of China, Qin Shi Huang (259–210 BCE), who unified China in 221 BCE. He established the Qin dynasty and is known for centralizing power, creating a unified script, building the Great Wall of China, and commissioning the Terracotta Army to protect his tomb in the afterlife.

There are claims that Qin Shi Huang was actually either an "Other" or assisted by "Others".[29] This would imply that "Others" had decided to invest in a second civilization, just in case.

Here is another of those wonderful twists of fate. In the same period, both Greece and China went through a period of philosophical development, in one case protected by a slave society and in the other by a tyrannical ruler, as well as virtual slaves known as Coolies.

We are now into the stages of "Other's" syllabus where the telepathic species have to make a much larger contribution. I can only point out coincidences and possibilities. Proof is impossible, probably to the delight of our skeptical friend, but there is clearly a coincidence here.

Chapter 3
THE LAST THOUSAND YEARS

a) THE RENAISSANCE

From when the Romans left northern Europe, it descended into virtually continuous warfare for almost a thousand years. The Vikings invaded, the Saxons invaded, the Normans invaded, and Britain fighting France was almost a national sport, only recently replaced by rugby football.

When we finally came to the Renaissance, things settled down a bit, and we started the development of art, literature and philosophy. It began in north Italy, in the 14th century, and spread north, coming to an end in Italy with the Sack of Rome (1527), and elsewhere in the 17th Century.

In Italy, it was fashionable to sponsor artists such as Leonardo Da Vinci, Raphael or Michelangelo, so it would only have required

"Others" to start such a fashion. Christopher Columbus managed to raise the funding for his expeditions.

The main cause of dissent in this period was the Roman Catholic Church, whose disapproval of the sciences caused Scientists to be labeled "Heretics" and, sometimes, as in the case of Bruno[30], this even resulted in their executions. Perhaps this was typical of human nature, to defend one's power, or perhaps it was instigated by a faction of "Others", who disagreed with the whole idea, fighting back.

The German Renaissance was inspired by the Italian Renaissance, with such famous personages as artists Dürer and Holbein, and the architects Wolf and Roritzer coming to the fore. Martin Luther had a major influence on the Protestant Reformation.

The Renaissance in Britain was inspired by the Italian and German Renaissances, with Shakespeare, Marlowe and Ben Johnson rising to prominence. The architects Inigo Jones and Christopher Wren flourished. King Henry VIII renounced the Catholic Church and became a Protestant. Explorers such as Drake and Raleigh obtained backing from Queen Elizabeth I. She survived a war with Spain and the Spanish Armada, but remained unmarried. The Renaissance here was brought to an end by the English Civil War in 1642, when extreme Protestants took to smashing church sculptures which they considered idolatry.

There were certain key movers over this period with Martin Luther standing out. He changed northern Europe and, subsequently, North America for ever. What was his source of inspiration?

b)   THE RAILWAYS

The world's first steam train to run on rails[31] was built for a bet by Richard Trevithick in 1804 and successfully hauled a load of iron on the Penydarren Tram-road in Wales. However, this locomotive was too heavy for the tracks. The first practical steam train for public use was Locomotion No. 1, built by Robert Stephenson and Company for the Stockton and Darlington Railway, which opened in 1825 and carried passengers and freight.

The Rainhill Trials[32], held in October 1829 for the Liverpool and Manchester Railway, was the first significant competition for steam locomotives to determine the best motive power for the new railway. The L&MR directors had not decided whether to use fixed engines with ropes or steam locomotives and resolved to hold trials to see if a steam locomotive would meet their requirements. George Stephenson, Robert's father, who worked for the L&M surveying the proposed route entered the Rocket, designed by his son Robert, which emerged as the winner, completing the trials and establishing the steam locomotive as the preferred choice for railway transport.

What made all the entries, apart from Stephenson's Rocket, fail to complete the tests? Was this an example of "Other" intervention?

There followed what came to be known as Railway Mania, which was a speculative boom and subsequent crash in the 1840s, primarily between 1844 and 1846, characterized by a frenzy of investment in railway companies and rapid, often unprofitable, railway construction[33]. Following earlier successful lines, investors poured money into thousands of new railway projects, leading to inflated share prices, but by late 1846, the bubble burst, resulting in financial ruin for many and a slump in the market.

Thus the British railway system was built by people who subsequently went bankrupt, enabling a small group of companies to buy up the various lines, and to form a viable operation. All this expertise travelled round the world, uniting countries as never before.

Was Railway Mania something initiated by "Others" to make the Earth seem smaller?

Much of the Railway Mania was attributed to the so-called Railway King, George Hudson[34], who launched scheme after scheme, eventually owning over 1000 miles (1,600 Km) of lines. He could sell anything, but could manage nothing. Who gave him the drive to do this?

c) <u>CHEMISTRY</u>

In Europe and Russia, the early days of Chemistry were the realm of the individual with use of a laboratory of their own. Some of the most famous were:

- **Antoine Lavoisier,** (1743-1794) "The Father of Chemistry" His most famous discovery was the law of conservation of mass, which states that whilst substances may change their state or form, they retain the same mass. He was also the first person to identify oxygen and hydrogen
- **John Dalton** (1766-1844) is these days remembered for proposing his atomic theory, which represents the basics of

chemistry today. He also studied and ultimately formulated a theory of atomic weight.
- **Amedeo Avogadro.** (1776-1856) His great discovery is what has come to be known as Avogadro's Law, which states that, under the same temperature and pressure conditions, the same volumes of different gases will have the same number of molecules.
- **Henry Cavendish**, (1731-1810) who discovered Hydrogen and proved that water was a compound, not an element, was intensely shy, and preferred to communicate by notes.

18th Century chemists received training primarily through apprenticeships, technical tours of mines and factories, and formal education at universities and specialized institutions like those for apothecaries.

The best training for a 19th-century chemist involved apprenticeships in busy shops or, increasingly, formal university education, with Germany eventually becoming the leading center for advanced chemical training by the mid-19th century. Early training relied on apprenticeships with a master, but this was often inadequate, leading to the development of specialized university laboratories and courses in Paris, Stockholm, and later, Giessen, Germany

So, why did Germany become the leading country for Chemistry?

18th Century Germany became prominent in chemistry training[35] due to strong state support, the development of a university-based research model emphasizing practical application, and the foundational contributions of prominent scientists like Stahl and

Klaproth. This laid the groundwork for Germany's later dominance in chemical research and industry.

Apparently, the "Others" would only have to exert influence on a few policy makers to initiate Earth's development in Chemistry.

It is also recorded that August Kekulé[36] who in the 1860s first identified that the Benzine molecule was ring-shaped, and Dimitri Mendeleev[37] who presented the first periodic table of the elements in 1869, both said that they first thought of the idea in a dream. Did someone place it there?

d) ELECTRICAL POWER GENERTION & TRANSMISSION

Belgian Born Zénobe Gramme is credited with inventing the first commercially successful Direct Current generator, the "Gramme dynamo," in 1869. His 1871 working model, featuring a new wire-wrapped ring rotor known as the "Gramme ring," produced significantly higher voltages than previous designs and demonstrated that the device could also function as a motor. This invention spurred the development of electric power for industrial applications.

The "first design" for an electric generator was a simple homopolar generator built by Michael Faraday in 1831 to demonstrate electromagnetic induction. In the United States, Charles F. Brush developed one of the first efficient and reliable dynamos (D/C generators) in 1876, and Thomas Edison's Pearl Street Station in 1882 was the world's first purpose-built central power station, using his D/C system to power incandescent lamps.

The genius Nikola Tesla was born in Smiljan in Croatia in 1856. His father was a priest but he wanted to study the sciences. He was educated in the Real Gymnasium in Carlstadt, Croatia, before going to read engineering at Graz Polyrechnic and Prague University. His main interest was electricity and he had a flash of inspiration in which he visualized totally the workings of an alternating current (A/C) motor.[38] Did he jump or was he pushed?

He went to work for the Continental Edison Company of Paris, where he gained a reputation as a highly talented Electrical Engineer. Unfortunately he was very naïve when it came to business, feeling he had been cheated out of a promised bonus. In 1884, he moved to New York, where he was invited to work for Edison directly. He continued developing his A/C ideas in his own time, but received no encouragement from Edison. Once again he was promised a bonus for solving a particular problem, but never received it when he succeeded.

Tesla and Edison parted ways after Edison's failure to acknowledge Tesla's contribution – something Edison was known for. He then set up with George Westinghouse, and began a campaign to demonstrate the superiority of A/C. Edison responded by initiating a smear campaign to prove A/C was dangerous and even used A/C in the electric chair to create the perception that A/C was deadly.

Tesla and Westinghouse's demonstrations and the inherent advantages of A/C for large-scale power distribution led to its adoption. A/C power became the standard for the world's electrical grid, establishing the future of electricity and eclipsing Edison's direct current system, forcing him to wind up his company.

Tesla took out patents on his inventions, and these were subsequently bought by Westinghouse in 1888. However, Westinghouse proved as ruthless as Edison, paying for the patents, but not the subsequent commission which Tesla had expected.

Once again, Tesla had proved to be a brilliant engineer, but unworldly in business matters. We will come back to Nikola Tesla soon.

e) TELECOMMUNICATIONS

In 1831, the year Faraday invented the D/C electrical generator, an American, Samuel Morse[39], heard about it and thought that a method of sending messages along wires was needed. He was a Professor of Fine Arts at New York University, yet he was the person to dream up the famous Morse code. It is amazing what the "Others" can make you dream!

Italian Guglielmo Marconi's early experiments in radio communication were financed by the British Post Office, a significant relationship that involved the first open sea wireless transmission[40] in 1897 at Lavernock Point and the acquisition of Marconi's coast station technology in 1909. The Post Office, led by its Chief Engineer Sir William Preece, championed and supported Marconi's groundbreaking work in wireless telegraphy, which laid the foundation for modern radio and wireless communication systems.

However, Tesla had patented his Tesla Coil and radio-tuning device six years before Marconi[41]. In 1893 he gave a lecture on how a radio should work, and patented his system in 1897, but never challenged Marconi.

This is a classic example of the perfectionist and the pragmatist. Tesla could have patented his idea earlier, but he waited until he had got it the way he wanted. Marconi publicized each stage in the development of his product, catching the public's imagination and succeeding in raising the funding he needed. He was the one who got the Nobel Prize in 1909, and is remembered as the inventor of the radio.

f) CHINESE DEVELOPMENT

The Renaissance in Europe was preceded in China by almost a thousand years. The Tang (618–907) and Song (960–1279) dynasties are often considered analogous "golden ages" for China's significant advancements in culture, science, and technology.

Key Chinese inventions that changed the world include the "Four Great Inventions": paper, gunpowder, the compass, and printing (woodblock and movable type). Other significant innovations include silk, the abacus, the wheelbarrow, the seismograph, and the God's gift to mankind - tea. These inventions profoundly impacted fields such as navigation, warfare, communication, education, and trade, fostering cultural exchange and driving global development.

All these inventions worked their way into Western Europe, for good or for bad, driving its culture forward. In return, China gained railways, chemistry, electrical power and communications. China was not interested in many of the products of Europe's Renaissance, preferring to develop its own artistic style.

Whilst this growth proved beneficial to both cultures, it led to competition between them, which sadly remains with us.

Chapter 4
THE LAST CENTURY

a) THE DEVELOPMENT OF US SOCIETY

In the US, in the 18th century, a national elite began to grow. This was often based on inherited wealth being invested in expanding technology and a subjugated population of slaves and exploited immigrants. Some managed to break into this elite by hard work on their part, and even harder work by their employees.

It has been said that, once you have made your fortune, you can afford to have scruples, and some of the most ruthless families in making their riches are now the most famous of philanthropists. Just ask any steelworkers from the last two centuries.

The likes of Edison and Westinghouse had both proved to be ruthless businessmen, determined to suppress innovations where they would cost them, and unlikely to honor promises which would cost them, unless forced to. To some, this seemed the way to the top.

This elite worked together to further their interests. In the 19th century they had significant influence on the government and, by the 20th century, they effectively had behind-the-scenes control of the country.

Take the Bush family. President George H W Bush had previously served as Vice-President to President Ronald Regan, and before then he was Director of the CIA and Ambassador to the United Nations. His son, George W Bush also became President. His father was an investment banker and a Senator. His maternal Grandfather was an investment banker.

Or take the Kennedy family. Joe Kennedy Senior[42] was ambassador to the United Kingdom before World War II. One of his sons was President Jack Kennedy, and of his other sons Robert Kennedy was a Senator and became Attorney General, and Ted Kennedy was a Senator. His father P J Kennedy was also a Senator. One of his grandsons R F Kennedy became Secretary of Health and Human Services under President Trump.

Tesla was like a sheep amongst a pack of wolves from the moment he arrived in America.

b)   POWER TRANSMISSION

Tesla was pretty wealthy once the battle of the currents was settled, although his future experimental projects were never going to be cheap. His first job was to oversee the construction of the Niagara Falls Power Station[43], and this enhanced his reputation. Then he struck out on his own, designing and patenting his Tesla Coil, and inventing and patenting radio years before Marconi.[44] He even built and publicly demonstrated a radio-controlled boat in 1897. He started working on the free distribution of wireless power.

In his notebooks from 1899, Tesla referred to 'stationary terrestrial waves' at approximately 7.5 Hz, a value strikingly similar to the Schumann Resonance that wasn't officially recognized until 1952. By energizing the earth-ionosphere cavity, he inadvertently tuned into the Earth's rhythmic pulse, years before scientific instruments validated its subtle presence.

Tesla needed money to develop his ideas, and he couldn't get a backer until, in 1902, J P Morgan[45], the driving force behind the General Electric Company (GEC), put up money in return for a 51% stake in the company holding the patents.

What Tesla didn't realize was that his wonderful ideas for free wireless distribution of power were a direct threat to GEC which had invested a fortune in their cable distribution system, and could wipe it out the way A/C had wiped out D/C and domestic gas.

At the first opportunity J P Morgan exercised his 51% controlling majority and, in 1903, stopped the work forever.

From then on, until his death in 1943, Tesla interested himself in many projects, was acclaimed as a genius, but faded into oblivion because his ideas seemed too fantastic at the time, even if they have often come to fruition much later, after his death.

He died in the Hotel New Yorker, and all his papers – 80 trunks full – were confiscated by the government. There were 163,911 documents yet Dr John G Trump from MIT, who was tasked to review them, was able to say - within 3 days – that there was nothing of military or security interest in them!

Yet when his papers were returned to his family, there were only 60 trunks full. The elite families had probably struck again – Dr John G Trump was the uncle of President Donald Trump. Tesla's ideas were not going to be used for the benefit of mankind but in the interests of the very rich minority, in a single country.

c)   EXTRA-TERRESTRIAL TECHNOLOGY

What the US elite most wanted were the power sources and the drive mechanisms of "Other" spacecraft. There were obvious differences between the drives of different species, as reported by witnesses[46]. Some gave off flames, and some gave off dangerous emissions, possibly radioactive or possibly electro-magnetic, which could cause burns, cancer or even death. It appeared that water was necessary for their operation.

Some "Others" had invisibility mechanisms, for their ships and portable. There are many descriptions of "Others" only visible from the waist up, or having no feet. This would be possible using such a device. Perhaps the portable power source for this device was limited. There are also reports of bullets passing through "Others" without harming them. Could this be the use of holographic imaging?

Many "Others" are described as levitating or hovering above the ground. This could be the result of portable technology, perhaps similar to the technology used in their spacecraft. Their non-lethal weaponry would also be useful, sending humans to sleep or controlling them.

Of course, the US elite were not the only ones after this technology. Russia and Nazi Germany were extremely interested as well.

Some of this technology would probably have been part of the "syllabus" of human education, but this had been interrupted.

d)   THE "OTHERS" TEAM

In Appendices 1 and 2, I have attempted to deduce which species were involved in the process of educating the Human race:

- **The Anunnaki**, possibly assisted by **Avians**, who first altered human genetics, and then taught them the basis of agriculture, mathematics, medicine, time measurement and many other things. They are reported to have moved on about 5,000 years ago.(See Appendix 3)
- **Dwarfs**, who possibly assisted humans in building monumental structures, and certainly taught them mining, and metal-working, both blacksmithing and metallurgy. They are reported to have withdrawn from human interaction in most places, and have returned to their earth colonies. (See Appendix 4)
- **The Pleiadians**, a telepathic tall Nordic race, who have probably been around as long as Dwarfs, and could have assisted in disseminating learning around the world.(See Appendix 5)

- **The Clarions**, a race of deceptively frail-looking humans with distinctive oriental almond-shaped eyes. These are telepaths who, because they did not stand out in a crowd, could have found it easier to approach those people who had to be influenced.(See Appendix 6)
- **The Tall Whites**, whose constellation of origin is unknown, but who probably oversee the whole project, and possible used their telepathy to influence human actions. They are still active all around the Earth. (See Appendix 7)
- **The Small Grays,** are responsible for defending Earth against attacks by malevolent "Others", and stopping anyone using atomic weapons. They also limited the power of the US elite, and later Majestic.(See Appendix 8)

This team had achieved so much before they came up against advanced human greed, which really put a spanner in the works.

e) AN ALTERNATIVE SYLLABUS

If universal free power was the next step in this syllabus, it appears to have gone seriously astray. It would seem that the "Others" then revised their tactics.

They arranged a serious of UAP crashes around the world, including bodies. There were crashes in:

- Russia   1941   Rostov[47]
- Italy   1933   Magenta[48]
- The United States of America   1947   Roswell[49]

There were probably many more around the world, including India, China and the United Kingdom. This meant that, even if the US elite had sole use of Wireless Power at first, they could not prevent "Other" technology from being available worldwide.

The United States went all-out to get as many crashed UAPs as possible, using its diplomatic might and threats of force if necessary.

- There was even a tense confrontation with Canada[50] on 12th February 2023 in Lake Huron, where the Canadian government elected to back down. It later turned out to be a Chinese spy balloon.
- On 25th August 1974, north of Presidio in Mexico, it is alleged that an armed US recovery team crossed the border[51] and claimed that all the Mexican soldiers there were already dead. They took the UAP and felt they had to burn everything else, bodies included, to avoid contamination.

The United States has a long-established program of reverse engineering[52] UAPs, with the intention of producing their own versions, and Russia and China probably do as well.

f) THE MAJESTIC TWELVE

The Roswell Crash occurred in 1947. That same year President Harry Truman signed the National Security Act which set up the Department of Defense, the USAF and the CIA. It is claimed that an organisation called Majestic-12[53] was set up as the same time, reporting directly to the President, to deal with all matters relating to UAPs.

In 1949, Majestic's secretary, James Forrestal, died after falling from a 16th-floor window at Bethesda Naval Hospital. While officially ruled a suicide, his death was shrouded in controversy and sparked numerous conspiracy theories about a Majestic power-grab.

It should be noted that a number of the members of this organisation were members of the elite families so, once again, they have a firm grasp on the levers of power.

Within a very short time, Majestic managed to sideline even the President, and to set up almost limitless funding from the State via black budgets and from industry acting in its own self-interest. The military have always resented political interference, choosing to forget that they have sworn to protect the nation as defined by its elected government. Perhaps they have chosen to re-define what is meant by "nation".

In his farewell address when leaving the White House, President Dwight Eisenhower[54] said:

"In the councils of government, we must guard against the acquisition of unwarranted influence, whether sought or unsought, by the military-industrial complex."

It is this military-industrial complex which has enabled Majestic to take control of the United States of America.

It is alleged that when President Carter took office, he was given an official briefing on UAPs and "Others" and that, afterwards he was found in his office, sobbing his eyes out[55].

The reverse engineering of UAPs has continued, probably now under the control of Majestic, and the US has now become a two part state.

NASA undertakes the public part of the space programme, whilst other agencies finance the private sector to undertake Majestic's space programme. NASA has to be very careful to avoid accidentally releasing anything about Majestic's activities. In these areas, the President has to do what he or she is told.

It has been reported that Majestic's operations are now so advanced that they could leave Earth behind if they wished[56]. One wonders why some billionaires are so keen to develop their own spaceships, and go to Mars.

It certainly would not be in the US elite's interests to cooperate with other nations, although they might work with other ruling elites. Otherwise, it would imply that they are working in the interest of the Human Race and, as I've already shown, this is most unlikely to be true. They would have to have shown a considerable change of heart since they pushed Tesla aside. They are still ruthless in dealing with any challenge to their power.

g) CURRENT UNITED STATES ACTIVITIES

Another genius who may have been used by "Others" was Dr Werner von Braun. This German was instrumental in the development of the first cruise missile – the V1 – and the first ballistic missile to reach space – the V2 – before moving to the United States to lead their NASA space program developing the massive Saturn V rocket which took men to the moon.

He clearly believed in "Others", and possibly that is where his inspiration came from.

The US has achieved far more in space than has been admitted. They have autonomous vehicles such as the X-37B[57], and crewed triangles such as the TR3B[58]. The hacker Gary McKinnon[59] showed that they have a number of craft, paid for by the American Taxpayer, and there are probably more under construction.

The most recent American spacecraft[60], that they admit to, include the Orion capsule for deep space missions, the SpaceX Dragon spacecraft for low Earth orbit transportation, and SpaceX's large-scale Starship system designed for lunar missions and beyond. The private Blue Origin spacecraft, such as New Shepard, are used for suborbital tourism, while other private ventures like the Intuitive Machines are conducting lunar surface missions. Northrop Grumman's newest, large cargo variant, the Cygnus XL spacecraft, also recently launched on a mission to the ISS.

To achieve newer spacecraft, the US has had to set up its own bases, secret from the Earth population and, possibly, from the various "Other" groups.

There is certainly one US base in Australia. Pine Gap near Alice Springs is claimed by the US to be a listening station set up to overhear electronic communications primarily in China and North Korea. However, chance sightings demonstrate that it is far more than that[61].

The Blue Mountains in Australia are claimed to hold a second base used for building and testing US spacecraft. In my opinion the book which makes this claim[62] derives far too many conclusions from too little data, combined with hearsay. There have been many UAP sightings in the area, but have these been "Other" or US spacecraft, or both?

There has been one interesting possibility to have come from Majestic's probable usurpation of Tesla's early attempts at a wireless power project, and the later discovery of the Schumann Cavity, possibly building on his early work. On 8th July 2003, the US launched the Opportunity Rover to Mars, where it started operating in January 2004. It was designed to operate for 90 days. It now holds the record for the longest-lasting Mars rover, operating for almost 15 years (5,498 Earth days).

Before it was launched, there were some last minute alterations to the power supply system (Not in itself unusual). Could it be that they had discovered that there was a wireless power supply system operating on Mars, left over from some "Other" activities, and they fitted a power receiver to Opportunity?

The supposedly solar-powered rover's mission ended in 2018 after a global dust storm, with NASA officially declaring the mission complete in February 2019.

There are reports of an operating Wireless Power system on earth in an underground cavern under Mount Denali in Alaska, which is reputed to contain what is known as the Dark Pyramid[63]. It has been claimed that this generates sufficient electricity to power the whole of Canada[64]. Is this an ancient facility or has the US built it? How did the US find it?

Tesla lost his patents in 1903, and in 1908 the Tunguska Incident occurred, where something large apparently exploded in mid air over

Russia. It has been suggested that it was shot down by copper "Cauldrons" spread out in Siberia, which are supposedly part of a defense system build by "Others" and powered by the Dark Pyramid in Alaska[65]. Scientists in the US, looking at Tesla's patents, might have detected it powering up beforehand, and realized what was happening. However, it is probable that Human technology was not sufficiently advanced at that time, but there have been several objects seen to break up over Russia since then, so they've had several more chances.

There have been claims of an ancient pyramid[66] found on Mars, and a number of claims that the Great Pyramid at Giza[67] was once a power generator for the Ancients, using granite monoliths as relays.

The "Others" have succeeded in finessing the US elite, but it will be interesting to see what happens next.

Chapter 5

THE WAY FORWARD

a) SPACEFLIGHT

You may remember that, back in Chapter 4, Section b, I described how it was found that the Schumann Cavity, the space between the ground and the ionosphere, could resonate so that it carried electrical power from the transmitter to anyone with the right receiver.

This may be how power is transmitted to all the "Other" craft flying over us, from Alaska and perhaps from as-yet undiscovered sites elsewhere. There would be no particular requirement for these craft to have inter-planetary capability, let alone inter-stellar. They could simply dock with mother-ships if this was necessary.

So, the various human nations have only been presented with examples of basically up-rated aircraft for them to reverse engineer. These do not use jet fuel and so are a step in the right direction.

Also, they effectively negate the US's temporary lead which they achieved by confiscating Tesla's papers, and keeping them secret.

I suspect that the US's claim, that they could up-sticks and leave Earth if necessary, was just about correct. They could only fly beyond the ionosphere using rocket fuel powered craft, but these would be very slow when compared with "Other" inter-planetary ships.

The "Others" have managed to regain control of Human education. However, it may be some time before the next stage arrives.

b)   TRADING WITH "OTHER" COLONIES

The various Earth nations would have had access to some of the "Other" technology on the crashed craft, such as their invisibility devices, and possibly their levitation devices, but the "Others" would have made sure there was no weaponry or mind-control equipment present.

However, most national governments would have some idea of where the "Other" colonies were in their country, even if they haven't told their general population. To avoid any confrontation due to resource limitation, they will probably enter into trade agreements with them. The colonies would still have the equipment they originally needed for construction, and they could either lease this out, or use it to undertake jobs for humans, in return for whatever supplies they needed.

It wouldn't be long before governments would be forced to admit what was going on, particularly if the "Others" announced that they had decided that "the time is right", as specified in the Small Gray's original contract[68] with the US.

It may be some time before the "Others" team decides to proceed to the next level in the "Syllabus". They would probably want to see all of this level's knowledge made available to the general population first.

What will come next?

APPENDICES

Earth's Alien Syllabus

APPENDIX 1 - The Little People   40
i   What are Little People?   40
     Table 1 Number of Little People in Argentina of different heights as estimated by witnesses   41
     Table 2 – Estimated heights of Little People in Argentina   43
ii   Various Books' Descriptions of Little People   43
     Table 3 Various books' descriptions of "other" physiology   44
     Note: The book by Neil Anami does not have page numbers.

APPENDIX 2 - Taller "Others"   45
     Table 1 Number of Taller "Others" of different heights sighted in Argentina as estimated by witnesses.   47
     Table 2 Species taller than 1.5m

APPENDIX 3 - Anunnaki   49

APPENDIX 4 - Dwarfs   52
i   Dwarfism   52
ii   Mythology   54
iii   The Dwarfs Today   55

APPENDIX 5 - Pleiadians   56

APPENDIX 6 - Clarians   59

APPENDIX 7 – Tall Whites   62

APPENDIX 8 - Small Grays   64

APPENDIX 1
The Little People

i. WHAT ARE LITTLE PEOPLE?

When someone is called small, what does that really mean? It could mean absolutely tiny, like a new-born babe, or it could mean a bit smaller than the speaker. This gives a range of about 0.3m (1') up to say 1.6m (5'3").

In an effort to understand how people think, I have looked at the complete list of "Other" sightings for Argentina from the book published by George Mitrovic[69], extracting those sightings of entities 1.4m or less in height. I chose this upper limit to avoid confusion with Small Grays. I have chosen Argentina because North America is dominated by Small Grays and Bigfoot (Yeti), and Europe seems to have a lot of different types of "Other" Little People.

Short	<0.5m	0.5m	0.6m	0.7m	0.8m	0.9m	1.0m	1.1m	1.2m	1.3m	1.4m
77	16	9	19	17	3	27	50	4	23	4	13

Table 1 Number of Little People in Argentina of different heights as estimated by witnesses.

Very small "Others" really only started being sighted in about 1965, and Small Grays, normally described as being 1.4m – 1.6m, only started abducting humans in South America on a big scale in about 1975.

It would appear that many people have difficulty in estimating heights without some sort of reference, whether their own height, or some convenient object. It is evident that over a third of the witnesses preferred to avoid the issue. So, some of these sightings of short people could be for entities up to 1.6m high, so I have had to exclude them.

Amongst the sightings, a half of those at 1.0m were green-skinned, and most at 0.6m were described as Cyclops. There were many sightings of really tiny Little People, less than 0.5m tall.

Assuming that all the real Little People are within the range 0.1m to 1.4m, I have looked at the various descriptions given in the sighting reports, trying to match them with their estimated heights where possible.

In Table 2 on the next page, I have tried to match the sightings with known species of "Others" where their height range fits. In Table 3, I have surveyed the various publications which attempt to describe these "Others", in search of commonality. Again, I have only considered those species described as shorter that 1.5m.

Together, Table 2 and Table 3 demonstrate a wide range of physiological variations in the Little People, and this implies that there are different species present rather than slight variations in one species. On this basis, there would appear to be no real reason why Dwarfs should not be considered as one more member of the Little People range of "Other" species, albeit about half with green skins.

Not all of these "Others" arrived on Earth at the same time. It is claimed that the species, called the Chaneques by the Mexicans, are the descendants of a spacecraft which crashed there thousands of years ago[70], whilst the 7 cm tall species, called the Jenglot by the Malaysians and Indonesians, only arrived in their spacecraft in the 1960s[71].

Bearded Brown hair brown or green skin	1.0m – 1.3m	Dwarf	
Bearded Red hair	1.0m – 1.3m	Dwarf	
Grey/Green Gnome	0.9m		
Manlike	0.2m	Chaneques	Mexico
Tiny Goggle Eyed	<0.1m	Jenglot	Malaysia Java
3 Horned	0.4m		
White Skin, Wings Young Boys/Girls	1.0m	Sprite	
Long Eared	0.6m		
Skin Like Acne	1.4m		
Triangular Body			
Large Headed	0.5m – 1.3m		
Green Horned	0.5m – 1.4m		
Square Headed		Ant People	
Hairy	0.7m		
Bald headed	1.2m		
Fangs & Claws	1.0m	Chupacabara	
Gray Large head & eyes	1.2m – 1.5m	Small Grey	

Table 2 – Estimated heights of Little People in Argentina

ii. VARIOUS BOOKS' DESCRIPTIONS OF LITTLE PEOPLE

This has proved an eye-opener. Out of 8 books, they can seldom reach agreement on any "Other" apart from the Small Grays, and even then it is only 5 out of 8 who do. Only Dwarfs and Arcturians get mentions by 3 out of 8 books, and then we are down to 3 species getting mentioned by 2 out of 8 books, and 9 species getting a single mention.

Earth's Alien Syllabus

Name	Height (m)	Pastore[72]	Spartacus[73]	Fredrich[74]	Anami[75]	McDaniel[76]	KGB[77]	Campobasso[78]	Sokol[79]	Details	Planet
Small Gray	0,9-1.3	77	4	-	?	-	-	157	63	Big Eyes	
Dwarf	1.0-1.3	-	-	33	?	27	-	-	-	Hairy	
Arcturians	1.0-1.3	-	74	19	-	-	-	-	14	Blue/green	
Langs	<0.70	258	-	-	-	-	45	-	-	Fairies	
Iguanoids	<1.0	-	-	-	?	-	-	231	-	Iguana	
Spotted Face	<1.0	-	-	-	-	55	-	258	-	Acne	
Sprite	<1.0	-	-	-	-	53	-	-	-	Wings	
Sylphans	1.3-2.0	-	-	-	-	-	-	-	156	Wings	
Batbazouls	0.7-1.3							257		Wings Ugly	
Goblin	<0.9	-	-	-	-	35	-	-	-	Big Ears	
Zeta Reticulan	1.3-1.9	-	-	-	-	-	-	175	229	Head & Eyes	
Little Green Men	<1.0	-	-	-	?	-	-	-	-	Green	
Ellina	Small	-	-	-	-	-	76	-	-	Elvish	
Dorsay	<0.5	-	-	-	-	-	39	-	-		
Small Reptoid	<1.0	-	-	-	-	-	-	253	-	Raptor-like	
Zeta Lizard Human	1.0-1.7	-	-	-	-	-	-	211	-	Large Head	

Table 3 Various books descriptions of "other" physiology
Note: The book by Neil Anami does not have page numbers.

It would appear that the Little People are of much less interest than "Others" that look like humans or are far, far taller. Nevertheless, it seems clear that Dwarfs and Arcturians seem the most common apart from Small Grays.

APPENDIX 2

TALLER "OTHERS"

Once again, I have tried to estimate the most likely of the taller "Others" to be present in Argentina[80] by looking at the frequency with which they have been sighted. There is a very wide range in heights, even when the Little People are excluded. There are a large number of sightings at 1.5m because this includes the Small Grays. On top of the sightings recorded here, there were 16 Yeti, 8 Cyclops, 7 Reptilians and 4 black Mothmen.

1.5m	1.6m	1.7m	1.8m	1.9m	2.0m	2.1m	2.2m	2.3m	2.4m
21	13	41	22	10	43	18	26	30	26

Table 1 Number of Taller "Others" of different heights sighted in Argentina as estimated by witnesses.

It appears that there are 3 peaks, at 1.7m, 2.0m and 2.3m. Hopefully these will coincide with the 3 species which are in the majority.

Looking at the same books which I used in Table 3 of Appendix 1, to identify the various species of Little People, I have listed, in Table 2 below, all the "Others" described, which are not shown as Little People. The names seem to be quite arbitrary – Take any star or creature and make it into an adjective. The KGB list seems only to have Small Grays in common with this list. Some descriptions are flatly contradictory.

With so many "Others" being described as tall Nordics or tall and human-like, you are left wondering how we ever knew that these were different species.

Earth's Alien Syllabus

Name	Height	Telepath	Description
Agathans	2.0-2.6m	N	Earth human (Telosians)
Alcyone Pleiadians	1.6-1.9m	Y	Nordic
Altarians	2.0-3.0m	Y	Human-like Blue/green/tan skin
Andromedans	1.7-2.1m	N	Blue skinned
Antarians	2.2-2.8m	Y	Human-like
Apunians	2.1-2.8m	Y	Nordic
Arcturians	1.3-1.6m	Y	Blue skin
	3.0-4.0m		
Arians	1.7-2.1m	Headgear	Nordic
Cassiopians	1.9-2.5m	Y	Human-like. Webbed fingers
Certans	1.9-2.5M	N	Human-like
Clarions	1.5-1.7m	Y	Petite human-like, slanted almond eyes
Cyclops	1.9-2.0m	Y	One eye
Cygnus Alpha	2.2-3.0m	Y	Tall human-like
Eridaneans	2.0-2.2m	Y	Nordic. Light blue skin
Itipurians	1.8-2.0	N	Thin humans
Lyrians	2.0-3.0m	Y	Feline
Klermers	2.5-3.2m	Y	Tall human-like
Koldashans	1.7-2.0m	Y	Human-like
Lady of Light	1.6m	N	Fair human. Skin glows
Lyrans	2.0-3.0m	Y	Human-like
Mantis	2.3-3.2m	Y	Insect. Triangular head
Melchizedeks	1.7-2.0m	N	Human-like
Mothman	2.0-2.6m	N	Dark fur, big wings.
Original Human Orions	3.3-3.8m	Y	Human-like
Pleidians	1.8-2.1m	Y	Nordic Fair graceful, slender
Procyonians	2.0-2.2m	Y	Feline
	1.8-2.8m	N	Human-like
Proxima Centaurians	1.6-1.7m	N	Human-like
Renegade Pleidians	2.6-4.2m	Y	Massive human-like
Reptoids	2.0-3.0m	Y	Scaly skin
Sagittarians	2.5-4.0m	Y	Human-like

Earth's Alien Syllabus

Name	Height	Telepath	Description
Saurians	1.8-2.2m	N	Lizard-like with tail
Sirians	2.0-2.7m	Y	Blue skin gills webbed hands
Soulzars	3.0-5.0m	Y	Hairless hybrids. Bald
Sylphons	1.2-1.8m	N	Gossamer wings
Tall Whites	2.1-3.1m	Y	Slender body
Thurbans	2.1-2.7m	Y	Reptilian
Titan Sirians	2.5-4.0m	Y	Blue skin, elongated head pointed chin
Unmites	2.5-2.8m	Y	Nordics
Vegans	1.8-2.2m	Y	Human-like
Venusians	2.0-2.8m	Y	Elegant, glowing skin (Valiant Thor)
	1.8-2.1m	Y	Human-like
Zeta Reticulans	1.0-1.6m	Y	Fragile body, big head & Eyes (Grays)

Table 2 Species taller than 1.5m

Based on these, the 1.7m peak could be caused by about 7 species but, only 3 of these are telepathic and many parts of the "syllabus" need a telepathic ability. This leaves Andromedans, Clarions and Melchizedeks. Clarions[81] are reported to have slanted oriental eyes, and these have been described in many places around the world, so these are probably the best bet.(eg Switzerland 1988)[82]

The wider peak between 1.9m and 2.2m points towards a Nordic of one form or another. The type whose height range best fits is the entity called a Pleiadian, and they are also telepathic. These are the claimed ancestors of many islanders in the Pacific Ocean as well as some Central American tribes.

The third broad peak between 2.3m and over 2.5m seems to belong to the Tall Whites, telepaths too.

APPENDIX 3

Anunnaki

The Anunnaki[83] are reputedly the "Other" species who first colonized Earth, perhaps over 50,000 years ago. They come from the planet Nibiru. It is claimed by some that this planet orbits our Sun on a vast elliptical orbit so that its appearances are very rare – every 3,600 years. However, I suspect that our astronomers would have found it by now, if that were the case.

They are described as very tall (perhaps 4-5 meters) humanoids, with the males having long beards, and they are often depicted in carvings as having wings.

They are claimed by Maestà Pastore[84] to have come to Earth in search of gold, supposedly to seed their atmosphere to create a greenhouse effect to keep their planet warm when it is distant from our sun. However, they decided to create a slave race to do the hard work.

Apparently speed was not of the essence, so they decided to modify the genes of a large primate, and then they sat back to wait for us to develop. I doubt if they remained on Earth whilst this had its effect, but they returned in about 10,000 BCE, to turn us into a self-sustaining workforce.

We are essentially someone's property.

They set up a colony in Sumeria, and started to teach our ancestors the basics of agriculture, medicine, animal husbandry, mathematics and astronomy. They must also have done this in the west of Central and South America, and in South Africa because these are areas where gold is plentiful.

In his book, Pastore gives an account of the Anunnaki politics which led to their colony being abandoned about 5,000 years ago, and we humans being apparently left to our own devices.

Earth's Alien Syllabus

APPENDIX 4

Dwarfs

i. Dwarfism

There are about 200 types of the condition known as dwarfism, and the birth rate in the United Kingdom is about 1 in 25,000[85]. In the past, without proper pre-natal care, this rate would have been much higher.

Of these births, 70% are what is termed "Achondroplasmia". There are 3 main visible types:

- Proportionate dwarfism where the individual is small but otherwise is perfectly normal.
- Disproportionate dwarfism where the longer bones of the body, particularly the legs, are much shorter and may the deformed.
- Seriously deformed people, perhaps even missing limbs, and often with mental deficiencies.

Back in history, these small people had to trade on their deformities if they could. There were troops of acrobats in Roman times, they worked as "Fools" to entertain the rich in the Middle Ages, and sometimes became their trusted advisors[86]. It has been wrongly suggested that people suffering from gigantism are more stupid than average. What is probably true is that people suffering from dwarfism seem to have more wit about them than average, given the proportion of them that rose to important positions. For example:

- Jeffrey Hudson, a small person born in 1619, became a trusted messenger for King Charles, and was eventually knighted.

- The famous Italian small person Bertholde became Prime Minister to the King of Lombardy.

What is missing is that you do not hear, today, of such small people being famed for their stone-working and metal-working skills, yet that is the tradition which is held in the folklore of many nations.

ii.   The Mythology

There is a clear distinction between the Small People who suffer from dwarfism, and the Dwarfs of mythology and extra-terrestrial visitations, who are a separate race, perhaps of "Others"

In his book, Claude Lecouteux[87] looks at the folklore of Europe, particularly Germany, France and Norway. He notes, in particular, that Dwarfs have been in northern Europe long enough to have joined the fringes of their pantheons, but that they appeared to have faded away in Victorian times, leaving the lore of Dwarfs working away in their secret smithys.

He describes how, in pre-Christian times, they were considered generally friendly, but the Church gradually demonized them:
- In France, Aubéron, who is described as King of the Dwarfs and a great magician, is noble and knightly in the manner of the Knights of the Round Table.
- In Germany, Alberich was father of the King of Lombardy and helps him win his bride. He may be model on which the story of Aubéron is based. However, he is also described as a masterthief.
- In Denmark, Dwarfs ruled the land of the dead.
- In Norway, the Dwarfs forged magical artifacts, and possessed hoards of gold

In Norse mythology Odin[88] the one-eyed Dwarf is the All-Father, the god of wisdom, war, magic, and death. He sacrificed his eye to drink from the well of Mimir, gaining vast knowledge. He also hanged himself from the World Tree, Yggdrasil, to learn the secrets of the runes. He is often depicted as a one-eyed, long-bearded old man.

In Spain there is the Duende[89], a mischievous spirit, similar to an elf or goblin, that lives in people's homes or in the wilderness. It can be a fierce protector of its dwelling and may bring good or bad luck. A benevolent Duende might perform helpful tasks or leave small gifts, while a displeased one could play pranks or cause disturbances.

The Mayans have a myth about a Dwarf[90]. He became king, then built a palace, and then a city which is part of the Uxmal ruins.

iii.   Dwarfs Today

Dwarfs' fame as miners comes from their diaspora when the Mediterranean tin became scarce and they spread out to find more, at the same time teaching the arts of mining and metal processing.

Since then, the mythology of Dwarfs appears to have faded, although their physical presence is still reported occasionally in sightings around the world, with evidence of at least two colonies[91], one in Northern Italy and one somewhere in Sweden. There are numerous sightings in Brazil and Argentina, suggesting that there is another one there, perhaps near Cape San Roque.

Anatolia, the source of early metal technology has no records or tales of Dwarf sightings in living memory[92].

APPENDIX 5

Pleiadians

THE PLEIADIANS[93]

As you might imagine, Pleiadians come from the Pleiades star cluster. A number of different "Other" species[94] are reputed to come from this cluster:

- **Renegade Pleiadians.** These are the tallest of the Pleiadians at 2.1m – 4.0m. They are greedy and malevolent, being basically self-centred.
- **Alcyone Pleiadians.** These are slightly shorter than the Pleiadians at 1.7m – 1.9m. They are generally kindly and friendly
- **Pleiadians.** These are typically humanoids who are slightly taller that humans (about 2m) and are muscular, fair skinned and blond haired[95]. For this reason they are also described as "Nordic".

They are fully telepathic, capable of conveying their thoughts amongst themselves, and with other species, whether they are telepathic or not.

Various cultures have mythical figures or groups of beings they associate with the Pleiades star cluster[96]:

- various Native American groups like the Kiowa and Nez Perce,
- in the Andes, where the Pleiades were seen as a symbol of abundance and were associated with the harvest season,
- in Hinduism where the Krittika group is identified with the Pleiades and

- in Hawaii, New Zealand (Māori), the Cook Islands, and Tuamotu,

Across Polynesia and beyond, the cultural significance of the Pleiades extends beyond a single island, serving as a celestial guide for navigation, agriculture, and ceremony.

Clearly, in the past, they were one of the groups who sought to teach races the first steps in civilization.

Even today, they often directly assist the more primitive tribes by offering healing, although they prefer not to become involved more than this.

APPENDIX 6

Clarions

CLARIONS[97]

Clarions are a humanoid race indistinguishable from humans, although they do have one marked characteristic – slanted almond shaped eyes that look vaguely oriental. They have petit facial features, but they do not stand out in a crowd, making them very useful in passing themselves as human.

However, there have been a number of occasions when a witness has commented on their eyes[98], suggesting that they are very much involved in furthering the "Other" syllabus. They are also telepathic, so they have been important influencers for many years.

Little is known about their way of life other than that they are a human extraterrestrial race, originating from the Aquila constellation, near a binary star. Some have been on Earth for 80 years, and is said that one of their bases is in the Amazon jungle.

They use disk-type UAPs, where the skin can become transparent, and it is said the ship is a living metallic creature.

Clarions first came to public awareness when they were implicated in a doomsday prediction for December 21st 1954 in the USA. Dorothy Martin[99] claimed to have received psychic messages from a planet called Clarion predicting flooding over much of North America and Europe. Some of the believers took significant actions that indicated a high degree of commitment to the prophecy. Some left or lost their jobs, neglected or ended their studies, ended relationships and friendships with non-believers, gave away money and/or disposed of possessions to prepare for their departure on a flying saucer, which they believed would rescue them and others in advance of the flood. I suppose they are still waiting.

We have no real evidence, apart from the word of Dorothy Martin, that Clarion was involved. I cannot quite see how this would have advanced the "Other" syllabus if they were, so I doubt that they had anything to do with it.

APPENDIX 7

Tall Whites

TALL WHITES[100].

There is some doubt about where these "Others" come from. Some say they are from Betelgeuse and are refugees on Earth[101], others say Arcturus[102] or from nearby

They are certainly tall, with heights ranging from 2m to 3m, with pale white skin, almond-shaped eyes, and are like tall thin humans. Their features are thin and angular, and they are very graceful. Despite their thin physique, they are very strong. They are telepathic, highly intelligent and live for centuries.

Their spacecraft use advanced anti-gravity propulsion and inter-dimensional travel, letting them move vast distances almost instantaneously.

They often carry devices which can manipulate electro-magnetic fields, and which can be used for personal protection. They are very skilled in healing.

They are generally thought to be here to monitor our technological advancement, and are considered by "Others" races to be capable mediators. They are reported to have had regular interactions with humans in a military and scientific context.

They have been on Earth for centuries, influencing the development of human cultures.

Earth's Alien Syllabus

APPENDIX 8

Small Grays

SMALL GRAYS[103]

These are the most easily identifiable of the "Other" species on Earth, being 1m - 1.5m tall with, as you might imagine, gray skin, They have very large heads, pointed chins, a small mouth and nose, and very big black eyes which dominate their faces. They have long thin arms and legs, and their hands can have 3 or 4 digits.

They are reported to come from Zeta Reticuli, and it is thought that they were a biological race, which has devolved into a part synthetic form. It is suggested that the genetic experiments, which they perform on humans, are directed towards recovering their lost biological selves.

They offered to assist the US government after a threatening show of force over Washington in 1952[104], by some unknown species – probably Small Grays! The entered into an agreement with the US, where they offered technology, in return for a base in the US, and permission to take animals and humans for study. This was agreed provided the number of humans was limited and they were returned unharmed.

They are reported to be unemotional and calculatingly cynical, and it wasn't long before they were ignoring the agreement when it suited them. They have been on earth for some time, and have managed an effective take-over of the United States, In particular, they are now responsible for most of the human abductions, performing operations on humans, particularly females, and mutilating animals. They initially acted solely in the US, but their activities are now spreading world-wide, with abductions reported in South America, North Africa, Pacific Islands and even New Zealand.

They are reputed to have caused the US, Russia and the UK to be unable to keep their nuclear weapons at readiness, by shutting down missile silos, and shooting down test missiles.

However, they are also responsible for protecting Earth from attack by hostile "Others". This is probably just a spin-off from protecting the "Other" colonies on earth, but we humans are along for the ride.

They have the agreed US base in Dulce, New Mexico, another near Catalina Island on the west coast of the US, one in Scandinavia and one somewhere in Europe.

Their abilities include telepathic control of humans, to facilitate their abductions.

INDEX

A/C 28, 29, 35
Abacus 31
Abductions 17, 76, 77
Agriculture 14, 20, 37, 60, 69
Alaska 44, 45, 46
Alberich 64
Almond-Shaped Eyes .. 38, 71, 73
Anatolia 16, 17, 20, 66
Anaximander 19
Ancient Mars Pyramid 45
Anti-Gravity 73
Anunnaki vi, 9, 11, 13, 14, 37, 59, 60, 61
Arcturians 53, 57
Aristotle 19
Aubéron 64
Avians 14, 37
Avogadro 26
Barcelona 17
Bases .... vii, viii, 18, 43
Basque 17
Ben Johnson 23
Benzine 27
Bertholde 64
Bessimer 18
Bigfoot 17, 50
Black Budgets 40
Blacksmiths 18
Blue Mountains 43
Blue Origin 43
Brain Development 11
Bronze Age 16, 17, 20

Bruno 23
Cape San Roque 65
Catalina Island 77
Cauldrons 44
Cavendish 26
Chaneques 51, 52
Chemistry .... v, 26, 27, 31
Christopher Columbus 23
Christopher Wren 23
Church 23, 64
Clarions 38, 57, 58, 70, 71
Clay Tablets 9
Cloud Formations 9
Colonies .... viii, 18, 38, 47, 65
Communications 31, 43
Compass 31
Confucius 20
Copper 16
Crashes 39
Curiosity 16
Cyclops 51, 56, 57
D allele 11
D/C 28, 30, 35
Dalton 26
Damascus steel 17
Dark Pyramid 44, 45
Denisovians 11
Diogenes 19
Dr John G Trump 36
Dr Werner von Braun 42
Drake 23
Duende 65

Dulce 77
Dürer 23
Dwarfism 63
Earthquake 10
Edison 28, 29, 33
Egyptians 14
Electrical Power 31, 46
Elite 33, 34, 36, 38, 39, 40, 42
English Civil War 24
Extra-Terrestrial .... v, 9
Fantasy 10
Faraday 28, 30
First Ballistic Missile 42
First Cruise Missile 42
Flores 11
Folklore 10, 13, 64
Gary McKinnon 42
GEC 35
George Hudson 25
George Stephenson 24
George W Bush 34
Great Pyramid at Giza 45
Greeks 14, 18
Green-Skinned 51
Gunpowder 31
Hales 18
Han Feizi 21
Hawaii 69
Heretics 23
Holbein 23
Holographic 37
Hominid 11
Hotel New Yorker 36
Hundred Schools of Thought ... 20

Inigo Jones 23
Intuitive Machines 43
Invisibility 36, 47
Ionosphere 46, 47
Iron 17, 18, 20, 24
J P Morgan 35
Jeffrey Hudson 63
Jenglot 51, 52
Joe Kennedy Senior 34
Kekulé 27
Klaproth 27
Lake Erie 18
Lake Huron 39
Laozi 20
Lavoisier 26
Leonardo Da Vinci 22
Levitation 37, 47
Little People .. v, vi, 49, 50, 51, 52, 54, 56
Magenta 39
Majestic .... v, viii, 38, 40, 41, 43
Marconi 30, 35
Marlowe 23
Mars 41, 44, 45
Marsh Gas 10
Martin Luther 23, 24
Mayans 14, 20, 65
Mencius 21
Mendeleev 27
Mesopotamia 13
Metallurgists 18
Meteor 10
Meteoritic Iron 17
Michelangelo 22

Microcephalin Gene 11	President George H W Bush 34
Military-Industrial Complex 41	President Harry Truman 40
Mind-Control 47	President Jack Kennedy 34
Miners 16, 18	President Ronald Regan 34
Morse 30	Presidio 39
Mother-Ships 46	Printing 31
Mothmen 56	Protestant 23, 24
Mount Denali 44	Pyramids 15, 79
Mozi 20	Pyrrho 19
Nan Madol 15	Pythagoras 19
NASA 41	Qin Shi Huang 21
Neanderthals 11	Railway Mania 25
Niagara Falls 34	Railways 31
Nibiru 60	Rainhill Trials 24
Nobel Prize 31	Raleigh 23
Nordics 56, 58	Raphael 22
Northrop Grumman 43	Renaissance .... v, 22, 31
Odin 65	Reverse Engineering 40, 41, 47
Officialdom 10	Richard Trevithick 24
Opportunity Rover 44	Robert Stephenson 24
Oriental 38, 58, 71	Rocket 24, 25
Orion 42	Rocket Fuel 47
Paper 31	Roritzer 23
Periodic Table 27	Rostov 39
Persians 14	Roswell 39, 40
Philanthropists 33	Sack of Rome 22
Pine Gap 43	Saturn V 42
Pittsburg 17	Schumann Cavity 43, 46
Plato 19	Schumann Resonance 35
Pleiades 38, 57, 59, 68, 69	Seismograph 31
President Jimmy Carter 41	Shakespeare 23
President Donald Trump 36	Silk 31
President Dwight Eisenhower. vii, 41	Simonside Dwarfs 18
	Skeptic 10, 11, 16, 20, 21

Slave 19, 60
Small Gray vii, viii, 17, 38, 48, 50, 52, 53, 56
Socrates 19
Spacecraft .. viii, 36, 37, 42, 43, 51
SpaceX 43
Stahl 27
Star Trek 15
Steam Train 24
Stone Age 16, 20
Stonehenge 15
Sumeria 9, 60
Syllabus .... v, 37, 39, 48, 58
Tall Whites 38, 58, 59
Tea 31
Telepathic . 21, 38, 58, 59, 68, 71, 73, 77
Tesla v, 28, 29, 30, 34, 35, 36, 42, 43, 44, 47

Tesla Coil 30, 35
Tin 16
Toledo 17
TR3B 42
Tunguska Incident 44
Underground vii
Underwater vii
V1 42
V2 .... 42, 79
Westinghouse 29, 33
Wheelbarrow 31
Wireless Power 39, 43, 44
Wolf 23
X-37B 42
Yeti 50, 56
Zeno 19
Zeta Reticuli 76
Zhuangzi 21

REFERENCES

[1] UFO Friend or Foe? by Martin Thomas Pub 2025
[2] We Own 29% - ET Has the Rest by Martin Thomas. Pub 2025
[3] We Own 29% - ET Has the Rest by Martin Thomas. Pub 2025
[4] Letters From Mesopotamia A L Oppenheim Pub University of Chicago Press 1967
[5] UFOs: Few answers at rare US Congressional hearing https://www.bbc.co.uk/news/world-us-canada-61474201
[6] Files released on 1974 'Welsh Roswell' https://www.bbc.co.uk/news/uk-wales-10863645
[7] The Order of the Dragon: The Battle Between "Other History and Accepted History by Colleen D Clements. Booksurge publishing 2006
[8] The Order of the Dragon: The Battle Between "Other History and Accepted History by Colleen D Clements. Booksurge publishing 2006
[9] A Humanized Version of Foxp2 Affects Cortico-Basal Ganglia Circuits in Mice by W Enard Cell Journal Volume 137, Issue 5p961-971May 29, 2009
[10] Denisovan and Neanderthal archaic introgression differentially impacted the genetics of complex traits in modern populations. Dora Koller et al BMC Biology 2022.
[11] Encyclopaedia of Alien Races by Màesta Pastore P151
[12] Letters From Mesopotamia A L Oppenheim Pub University of Chicago Press 1967
[13] Bird-Headed Deity, Denver Art Museum web site
[14] Pyramids and Ziggurats – National Geographical Society Education
[15] The Chinese Pyramids and the Sun, AC Sparavigna · Torino Polytechnic 2012
[16] What's Inside the Pyramid at Chichén Itzá? Britannica, 2025
[17] Bronze Age Source of Tin Discovered The University of Chicago Chronicle 1994
[18] Bronze Age Source of Tin Discovered The University of Chicago Chronicle 1994
[19] Magic, Myth & Mystery of Dwarfs by Virginia Hagan Pub Cherry Lake USA 2019
[20] Iron Age, historical technological and cultural stage Brirannica 2025
[21] First evidence of crucible steel production in Medieval Anatolia Science Direct 2022
[22] Toledo, Spain, has been a sword-making hotbed for 2,500 years — now just 2 artisans are keeping the tradition alive. Business Insider 2021
[23] UFOs, Humanoids and Strange Phenomna of Andorra, Gibraltar, Spain & Portugal by George Mitrovic. Pub Dunstable, UK.
[24] Amazing Encounters with UFOs of East Coast America V2 by George Mitrovic

Dunstable.
[25] The Duergar, The Dwarves of Simonside Hills. Spookyisles.com
[26] UFOs, Humanoids and Strange Phenomena of England by George Mitrovic.
[27] Bessemer Process Britannica 2024
[28] Hundred Schools Chinese History Britannica 2017
[29] Inf News/ history Did Aliens Help Qin Shi Huang to Rule the World?
[30] Giordano Bruno Italian philosopher Britannia 2025
[31] Richard Trevithick English Engineer Britannia 2024
[32] The Rainhill Trials by Anthony Dawson Pub 2018
[33] Deriving the Railway Mania. G Campbell Queens University Belfast Portal 2013
[34] George Hudson, the Railway King by Matthew Wells. Pub Pen& Sword 2024
[35] Social Support for Chemistry in Germany by Karl Hufbauer, University of California Press 1971
[36] Image & Reality by Alan Rocke Pub Amazon 2010
[37] Discovery of the Periodic Table by Matthew Lyons History Today 2021
[38] The Man Who Invented the Twentieth Century P26, by Robert Lomas. Headline Book Publishing 1999.
[39] Samuel F. B. Morse by Greg Timmons Biography>Famous Inventors>Famous Painters 2019
[40] Marconi's first radio broadcast made 125 years ago by Jonathan Holmes Pub BBC News/England/Local News/ Somerset 2022
[41] The Man Who Invented the Twentieth Century P261, by Robert Lomas. Headline Book Publishing 1999.
[42] Joe Kennedy: A Complex and Shocking Ambassador by Susan Ronald, Pub MacMillan 2021
[43] The First Hydro-Electric Power Plant in The World The Tesla society Website I
[44] The Man Who Invented the Twentieth Century P261, by Robert Lomas. Headline Book Publishing 1999.
[45] Nikola Tesla and the Tower That Became His 'Million Dollar Folly' by Gilbert King Smithsonian Magazine 2013
[46] UFOs, Humanoids and Strange Phenomena of Argentina, Chile, Paraguay and Uruguay by George Mitrovic. Pub Dunstable, UK
[47] UFOs, Humanoids and Strange Phenomena of Russia P34 by George Mitrovic Pub Dunstable UK
[48] Pentagon 'whistleblower' claims Vatican helped US retrieve UFO from Benito Mussolini New York Post June 13 2023

Earth's Alien Syllabus

[49] The Day After Roswell by Corso & Birnes Amazon 2017
[50] Unidentified object' downed by U.S. fighter jets over Lake Huron PBS News Nation Feb 2023
[51] Magic Eyes Only p278 by Ryan Wood. Pub Wood Enterprises USA 2024
[52] Dreamland: An Autobiography by Bob Lazar Pub Amazon 2021
[53] The Alien Colonisation of Earth's Waterways P181 by Debbie Ziegelmeyer, Pub UnX Media USA 2021
[54] President Dwight D. Eisenhower's Farewell Address (1961) US National Archives
[55] The Secret Space Program and Break-Away Civilisation by Richard Dolan P18
[56] The Secret Space Program and Break-Away Civilisation by Richard Dolan
[57] Boeing Defense Autonomous Systems
[58] https://www.msn.com/en-au/news/other/top-secret-anti-gravity-spy-plane - tr3b-black-manta/vi-AA1w0LhM
[59] Cibernews: Tech 2023
[60] Spacecraft Type List Spacecraft.Fandom.com
[61] UFO Friend or Foe? P42 by Martin Thomas Pub 2025
[62] Blue MountainsTriangle by Rex & Heather Gilroy, URU Publications 2006
[63] The Dark Pyramid and Violent Nature www.imdb.com/title
[64] Pyramid in AlaskaCan Power All of Canada? You-Tube: History 22 Aug 2023
[65] UFO Friend or Foe? P79 by Martin Thomas Pub 2025
[66] The Enigma of the Mars Pyramid by Boris Bigalke Books on Demand 2024
[67] Pyramid Power by Toth & Nielson Destiny Books 1999.
[68] The Alien Colonisation of Earth's Waterways P184 by Debbie Ziegelmeyer, Pub UnX Media USA 2021
[69] UFOs, Humanoids and Strange Phenomena of Argentina, Chile, Paraguay and Uruguay by George Mitrovic. Pub Dunstable, UK
[70] We Own 29% - ET Has the Rest by Martin Thomas. Pub 2025
[71] We Own 29% - ET Has the Rest by Martin Thomas. Pub 2025
[72] Encyclopaedia of Alien Races by Maesta Pastore Pub Amazon
[73] UFOs & Aliens 2 by Carl Spartacus. Pub Dunstable UK
[74] Alien Races by Alan Fredrich 2021
[75] Alien Species Book by Neil Anami Pun Dunstable UK
[76] Illustrated Guide to Reported Alien Species by David McDaniel
[77] The Secret KGB Book of Alien Races Pub Dunstable UK 2025
[78] The Extraterrestrial Species Almanac by Craig Campobasso 2021
[79] The Alien Archive – The Ultimate Alien Database Jacob Sokol 2025

Earth's Alien Syllabus

[80] UFOs, Humanoids and Strange Phenomena of Argentina, Chile, Paraguay and Uruguay by George Mitrovic. Pub Dunstable, UK
[81] The Extraterrestrial Species Almanac P47 by Craig Campobasso 2021
[82] UFOs, and Strange Phenomena of Austria, Belgium, Estonia, Germany, Holland, Kaliningrad, Latvia, Lithuania, Luxembourg, Poland and Switzerland p245. By George Mitrovich
[83] The Extraterrestrial Species Almanac P23 by Craig Campobasso 2021
[84] Encyclopaedia of Alien Races PP 146-163 by Maesta Pastore Pub Amazon
[85] The Natural History of the Dwarf by Richard Carrington Cosmo Books, Wem, England. Article from The Saturday Book, London 1958.
[86] The Natural History of the Dwarf by Richard Carrington Cosmo Books, Wem, England. Article from The Saturday Book, London 1958.
[87] The Hidden History of Elves & Dwarfs by Claude Lecouteux Pub Inner Traditions 2018
[88] The Origins of Wizards. Witches and Fairies. Simon Webb Pub Pen & Sword History 2022
[89] About Duende   duendedrama.org
[90] Magic, Myth & Mystery of Dwarfs by Virginia Hagan Pub Cherry Lake USA 2019
[91] We Own 29% - ET Has the Rest by Martin Thomas. Pub 2025
[92] UFOs, Humanoids & Strange Phenomena of Africa, Asia & the Middle East pp300-309 by George Mitrovic. Dunstable Uk
[93] The Alien Archive P123 – The Ultimate Alien Database Jacob Sokol 2025
[94] The Extraterrestrial Species Almanac pp101-109 by Craig Campobasso 2021
[95] The Extraterrestrial Species Almanac P103 by Craig Campobasso 2021
[96] Close Encounters of the Hawaii Kine. Hawaiian Airlines Issue 28-3
[97] Extraterrestrial Species Almanac P47 by Craig Campobasso 2021
[98] UFOs, Humanoids and Strange Phenomena of Austria, Belgium, Estonia, Germany, Holland, Kalingrad, Latvia, Lithuania, Luxembourg, Poland and Switzerland. P 245 (1988) By George Mitrovic Dunstable
[99] When Prophecy Fails by Festinger et. al. Pub Harper Torch 1956
[100] The Alien Archive P161 – The Ultimate Alien Database Jacob Sokol 2025
[101] Encyclopaedia of Alien Races PP 98 by Maesta Pastore Pub: Amazon
[102] Extraterrestrial Species Almanac P29 by Craig Campobasso 2021
[103] The Alien Archive P63 – The Ultimate Alien Database Jacob Sokol 2025
[104] UFOs in US Airspace Hard Evidence P403 by John Scott Chace USA 2020

www.ingramcontent.com/pod-product-compliance
Lightning Source LLC
Chambersburg PA
CBHW061224070526

44584CB00029B/3972